About the Author

Abhishek Talwar (AIEMA) is a certified environmentalist and author. He has the gift of narrating difficult concepts with simplicity and fun. This ease of narrative combined with the values of love for the environment, loyalty and friendship that shine through each story make these books truly entertaining and enriching for children.

About the Illustrator

Sonal Goyal completed her masters in fine arts from College of Art, Delhi. Having been creative head at a leading publishing house, she is now an independent artist. Her love for drawing and books has led her to creating cute and lovable illustrations that spread smiles.

Published by Ritika Talwar
All Copyrights and Trademarks: Ritika Talwar 2020
All rights reserved
ISBN 978-81-952948-5-5
Typeset in Sabon LT Std Roman

www.biplobworld.com

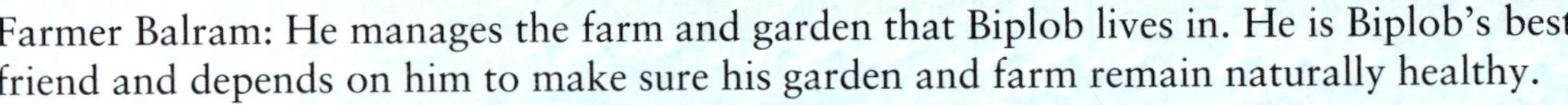

Biplob the Bumblebee: The eco – warrior superhero is the cleverest bee. Ever. He uses only his wits to solve the most difficult problems in the most natural, eco – friendly manner!

Farmer Balram: He manages the farm and garden that Biplob lives in. He is Biplob's best friend and depends on him to make sure his garden and farm remain naturally healthy.

Spot the Leopard: This brave leopard lives in the forests near Balram's farm. Since Biplob saved his life, he is loyal to the eco – warrior bee and always ready to help.

Addy & Avantika: Two siblings who live in the city and visit Biplob's garden whenever their school has holidays.

Karini - She is the matriarch head of the elephants. Short – tempered and fiercely protective of her brood. She is intelligent as only an elephant can be.

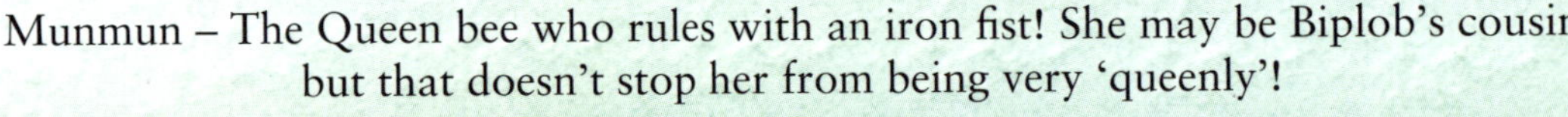

Munmun – The Queen bee who rules with an iron fist! She may be Biplob's cousin but that doesn't stop her from being very 'queenly'!

1. The Train – Boon or Bane?

"They look amazing!" whispered Aditya breathlessly as they crouched behind the bushes.

"Yes, aren't they regal?" said Biplob, admiring the herd of elephants in the distance.

"The baby elephant is so cute, walking between his mama's legs," cooed Avantika.

"That's called a calf and the mother elephant is called a cow," corrected Biplob.

They stayed hidden, watching as the elephants crossed a stream and went into the jungle beyond.

As the three headed back, chattering excitedly, they met
farmer Balram. "There you are!" he remarked. "I've been looking all
over for you. Never mind, hurry along before we're late for the inauguration."

"I almost forgot!" exclaimed Biplob as they started following him, the new railway
station's inauguration is today. Hurry up or we'll miss the fun."

"I hope they have samosas," said Addy as Avantika rolled her eyes.

The village headman cut a ribbon and broke a coconut to inaugurate the spanking new railway station. Farmer Balram flagged off the first train with aplomb as everyone applauded.

"Now we can easily send all our farm produce to the city," beamed the headman as the train chugged out of the station. Everyone relished the delicious jalebis and samosas, returning home happy as can be.

Next morning, Addy and Avantika woke to find farmer Balram and Biplob looking extremely gloomy.

"What happened?" asked Avantika.

He silently showed them the newspaper. They were aghast to see the news – the train had run over an elephant the previous night.

"Oh no! This is horrid!" cried out Addy as Avantika burst into tears.

"I hope it wasn't the mother of that cute little calf we saw," she sobbed.

"Why didn't the train driver stop?" asked Addy.

"He tried," replied farmer Balram, "But it was too dark and the track cuts across an elephant corridor. The train couldn't stop in time."

Addy was confused. "But there's no corridor in the jungle. We saw it ourselves yesterday."

"Elephants need so much food that if they stay at one place for too long, they'll eat up the entire forest," explained Biplob. "So, they move over long distances in search of food, following fixed paths called 'elephant corridors'.

Humans think it's just a jungle but for the elephants, it's as clear as a highway.
Unfortunately, the train tracks cut across this elephant corridor."

"So more elephants could die,"
sobbed Avantika inconsolably.

"Not if we can help it," said farmer Balram grimly.
"We're meeting the railway officers to find a solution."

They impatiently waited for farmer Balram to return from his meeting. No games held their interest and even Floppy couldn't entertain them with his antics. Finally, farmer Balram returned. Biplob took one look at his face and said, "It seems the news isn't good."

"That's right," he replied, "The railway officers were sympathetic but have no idea how to prevent future accidents. We cannot even move the tracks to go around the elephant corridor as that will cost too much and add 10 hours to the journey!"

Poor Biplob got nervous and started buzzing faster. "Arrghh! Get away! Your buzzing is driving me crazy!" yelled the elephant. Biplob hurriedly buzzed away.

"Any luck?" asked farmer Balram when Biplob returned. The bee shook his head sadly.

"Well, at least you tried. Why don't you go visit your cousin sister's hive? It'll take your mind off these troubles," suggested farmer Balram, reassuringly.

Biplob nodded, "You're right. I've heard Munmun is now the queen of her hive. Maybe the break will give me an idea on how to save the elephants."

When Biplob reached the hive, he found thousands of bees busy making honey! "Hello, I'm here to see Munmun." All the bees stopped working and looked shocked.

"He dares to call her Royal Buzziness by name?" whispered one. "Maybe he's new around here," remarked another. "What do you want with her Royal Buzziness?" asked a bee, buzzing out importantly.

'Must be the supervisor,' thought Biplob.

"I'm Munmun's cousin, Biplob, here to visit her," explained Biplob.

"Cousin, eh? You look nothing like her Royal Buzziness. Anyway, if you say, you must be her cousin," shrugged the bee.

"Of course I am!" insisted Biplob indignantly, "And what is this Royal Buzziness…I mean, business?"

"Please refer to her Royal Buzziness with respect," scolded the bee haughtily. "Come on, let's see if she gives you an audience."

Biplob buzzed along behind him. Deep in the hive, they found a very plump bee on a throne, a crown on her head.

"Munmun, is that you?" Biplob asked, peering at her.

"Forgive me your Royal Buzziness. This fellow claims to be your cousin. Should I knock him on his head for calling you by your name?" interrupted the bee. "That isn't necessary Tina. Leave us," replied Munmun, looking at Biplob.

"You look different! Come closer," Munmun ordered Biplob.

Just then Tina returned, breathless. "Your Royal Buzziness, the warriors have returned!"

"And?" asked Munmun.

Tina puffed up with pride. "They've driven 120 elephants from the valley of flowers. We can grow the new hive without worrying about those pesky giants trampling any plants!"

"Your bees can drive away elephants?" asked Biplob, amazed.

"You – speak only when her Royal Buzziness asks a question," scolded Tina.

"That's alright Tina. Biplob is yet to learn our rules," said Munmun. Turning to Biplob, she explained, "Elephants cannot stand our buzzing. They hear the buzzing of a thousand bees and run in the other direction."

"That's strange," said Biplob. "I thought their skin is so thick that even our sting doesn't hurt them."

Munmun nodded. "Maybe, but this is how it is."

Biplob's face lit up. "Munmun, I need your help. Could you set up another hive near our farm?"

Munmun tittered, "Help you? I've only tolerated you because you're my cousin." She turned to Tina. "Imprison Biplob till he learns how to speak to his queen."

Tina and four guards gleefully grabbed Biplob.

"I'm your cousin," screamed Biplob. "You cannot do this!"

But Munmun was already dozing off…

2
Warriors
in the Sky

After Biplob was in prison for two days, Munmun sent Tina to fetch him.

"The prisoner's here your Royal Buzziness," Tina announced to a dozing Munmun, who sputtered and opened her eyes.

"Eh, what? Oh yes, Biplob. So, have you gotten off your silly notion of living in a garden? Will you live in the hive amongst the other bees?" she asked.

But Biplob would have none of it. "You already know my answer, Munmun. I will never abandon my friends for anything."

Munmun just nodded shrewdly and instructed Tina, "Take him back to prison. Let him stay without food for three days. Maybe that will bring him to his senses."

As Tina grabbed Biplob, two injured bees buzzed in.

"We have been attacked!" gasped one bee.

"What?" exclaimed Munmun.

"There were five Giant Hornets your Royal Buzziness. We tried thermo-balling but it was no use. They captured 150 bees!" sobbed the other one.

"Impossible," thundered Munmun, "You must not have done the thermo-balling properly. Double the guard and ensure they are taught a lesson when they return."

Back in prison, Biplob asked, "What is thermo-balling?"

Tina sneered, "That's what comes from not living in a hive. You know nothing! Our stings are ineffective on the hornet's hard skin but they cannot withstand too much heat. So, her Royal Buzziness invented thermo-balling. Hundreds of us buzz around the hornet, creating so much heat that it collapses!"

"Wonder why it didn't work this time," said Biplob thoughtfully.
Tina just scowled.

Biplob was famished, surviving only on water. By nightfall, he was very hungry, sad and listless.

However, next morning, he woke up feeling surprisingly stronger and better than the previous day!

Suddenly, it struck him. "Quick, I must see Munmun immediately!" he shouted at the guard.

Thinking that Biplob had decided to join the hive, the guards took him to Munmun.

Surprisingly they found Munmun wide awake, hovering above her throne. Finally, her poor wings gave away under her weight and she plonked back down.

"Why do you look so worried?" asked Biplob. "Thermo-balling isn't stopping these hornets," replied Munmun with worry, before sharply looking at Biplob.

"Enough of that. Have you come to apologize and accept me as your queen? Has staying hungry taught you your lesson?" she asked.

"Yes," replied Biplob. "Just not the lesson you think."
Munmun looked puzzled.
"I've learnt why thermo-balling isn't working any longer." Biplob stated.
"Huh? Yesterday you didn't even know what it is," interrupted Tina.
"Quiet Tina," snapped Munmun. "Biplob was the smartest bee even when we were children."

"The hornets have gotten used to thermo-balling," continued Biplob, "Yesterday, without food I felt tired and weak. I had never experienced such hunger. But today, my urge to eat was not as strong as last night. I had some water and was fine. I realized you can get used to anything, even hunger! Similarly, the hornets have gotten used to the heat generated by thermo-balling. Now they just wait for you to tire out before attacking."

Munmun reflected on these words and asked, "Do you know how to stop them?"

Biplob nodded, "Yes but I'll tell you how on one condition."

"You want to be king of the hive?" asked Munmun suspiciously.

Biplob looked shocked. "Of course not! Every hive only has a queen, never a king. Is that why you've imprisoned me, thinking that I've come here to become king?"

Munmun looked down, ashamed.

"I want you to send 2,000 bees with me to set up a new hive near my home," said Biplob.

Munmun smirked, "Ha! So you want to be king of your own hive and want my bees for your subjects."

Biplob sighed, "No, Munmun. You'll be queen of both hives. I only want your bees as an ELEPHANT ALARM."

"Elephant alarm?" asked Munmun.

Biplob explained how the elephants near his village were endangered by the train. "If the elephants are looking to cross at the same time as the train, your bees will buzz loudly and drive them away. At other times when it's safe to cross, they'll leave them alone," he concluded.

Munmun thought carefully before nodding, "I agree. Now, how do we stop these hornets?"

3

The buzz of success

"Hornets breathe through their stomachs. When they inhale, their stomach expands and the hard scales covering their body move apart for air to enter. When they exhale, the scales close again," explained Biplob, his eyes twinkling.

"Instead of thermo-balling, we swarm them so closely that they are unable to expand their stomachs to breathe! Without breathing, they'll collapse."

Munmun nodded, "That may work. After all, you can't get used to not breathing!"

The rest of the bees too were relieved. Finally, they they had a plan to stop the hornets!

When they saw the aggressive hornets approaching, the bees all got ready. As soon as a hornet came close, they swarmed it. The hornet just laughed, "Your thermo-balling cannot stop me!"

The hornet was in for a shock though.

Instead of thermo-balling, the bees swarmed
him closely until he couldn't part his
stomach scales to breathe! He grew weak and faint.

"Please let me go! I promise we'll never attack again," he gasped.
"And return all the bees you've taken prisoner?"
demanded Biplob.

"Yes, yes, I promise," gasped the hornet, barely conscious.

Biplob signalled and the bees flew apart, allowing it to
breathe again.

The hornet kept his word and released all the prisoner bees. Munmun was beside herself with joy. "My dear cousin, you've saved my hive. Come, sit with me on the throne!" she beamed at Biplob "Thank you Munmun but the throne is yours. I only hope you remember your promise to me," said Biplob.

She simply smiled and hugged Biplob, tears of happiness streaming down her face.

Munmun clapped her hands, "Tina, take 2,000 bees with Biplob to set up a new hive – the 'Trumpeteers' Haven'. Not a single elephant should be hurt by a train ever again."

Turning back to Biplob, she said, "Remember, this is your home too. Promise me that you will visit again soon."

Biplob laughed and agreed, happy to be friends with his cousin once again.

Balram was shocked when Biplob returned with so many bees!

"Let's find the best place for the beehive so they can drive the elephants away when a train approaches," he gushed when Biplob narrated to him all that had happened.

Soon, they found the perfect spot next to the tracks. In no time, the new hive was buzzing with activity. Everyone was busy as only bees can be.

Later that evening, Balram and Biplob
hid near the tracks, watching as a train approached
in the distance. Sure enough, the herd of elephants ambled
along just then, looking to cross the tracks. However,
the bees at 'Trumpeteers' Haven' were alert as ever!

"Get ready," announced Tina. "Let's ensure no elephants
reach the tracks until the train has passed by safely."

All the bees flapped their little wings furiously, making a loud buzzing sound.
"NOW!" shouted Tina and they all flew towards the elephants, causing the matriarch to
trumpet loudly before turning around and running away, her herd in tow.
The train crossed the corridor safely; not a single elephant was hurt.
"It worked splendidly!" beamed farmer Balram. "Now the elephants
will never face any danger from the train."

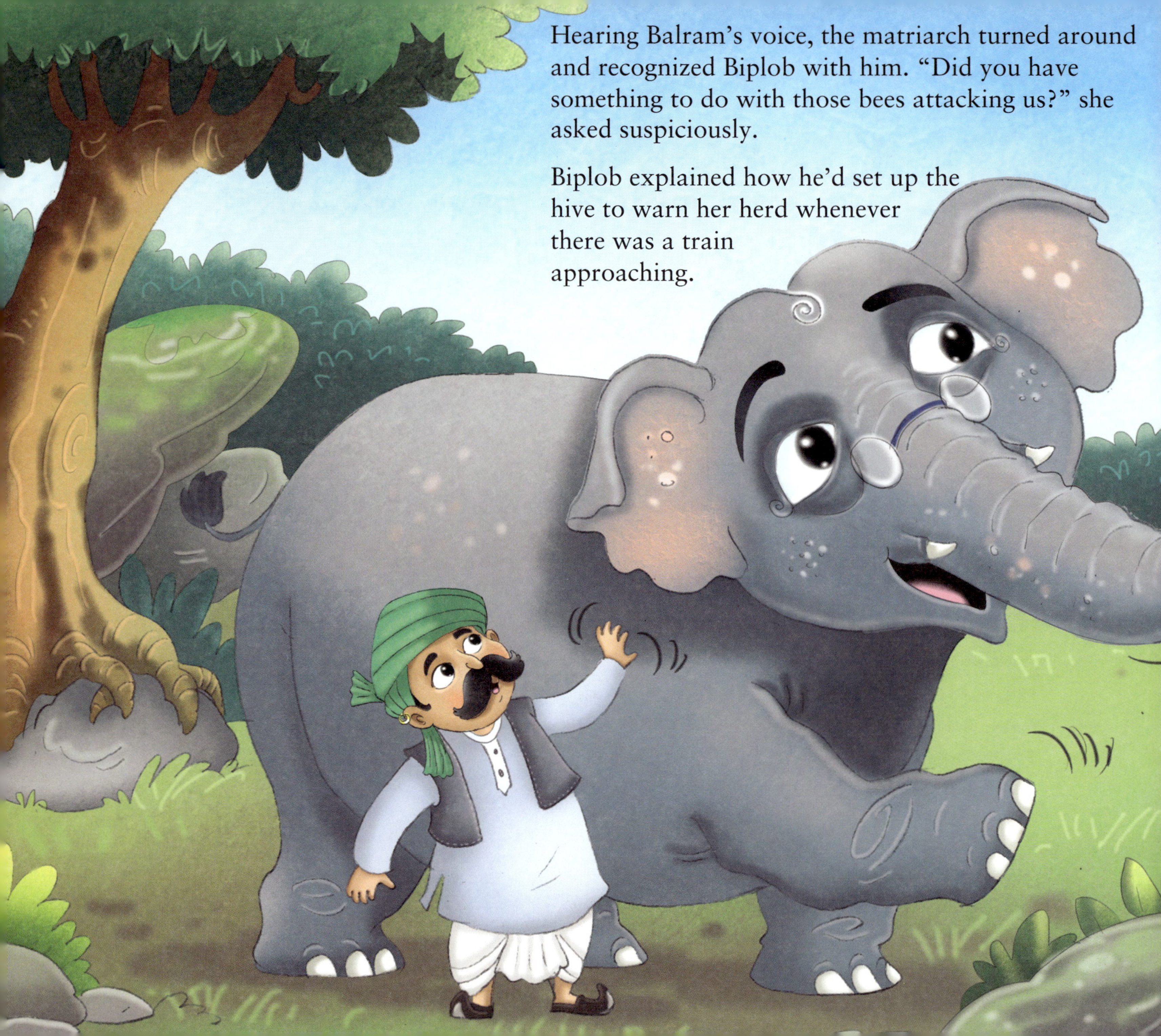

Hearing Balram's voice, the matriarch turned around and recognized Biplob with him. "Did you have something to do with those bees attacking us?" she asked suspiciously.

Biplob explained how he'd set up the hive to warn her herd whenever there was a train approaching.

She looked amazed. "That is brilliant! You are
a true friend of the jungle and all who live in it.
I, Karini, promise that if you ever need our help, we'll
always be there." Biplob humbly bowed to Karini as she
left with her herd.

Farmer Balram proudly smiled at Biplob, as he happily buzzed
with joy. After all, there was much to celebrate!

BIPLOB THE BUMBLEBEE
Book Series

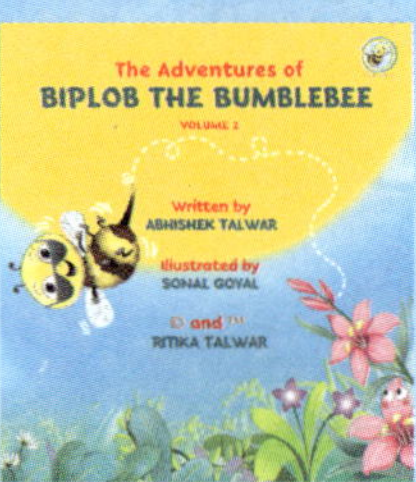
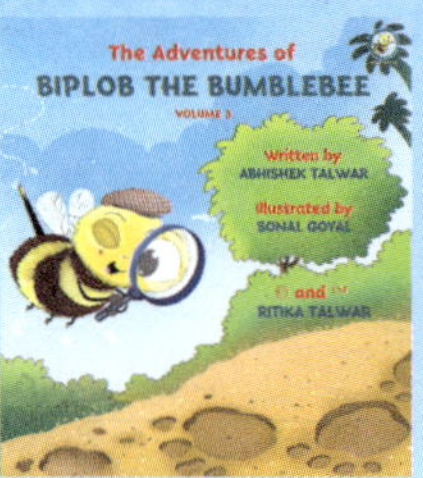

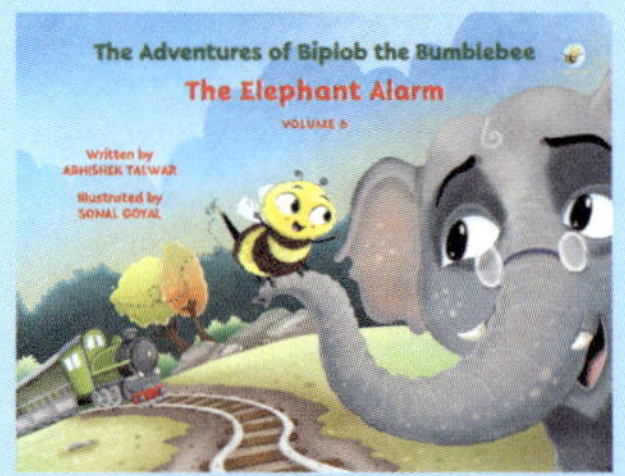
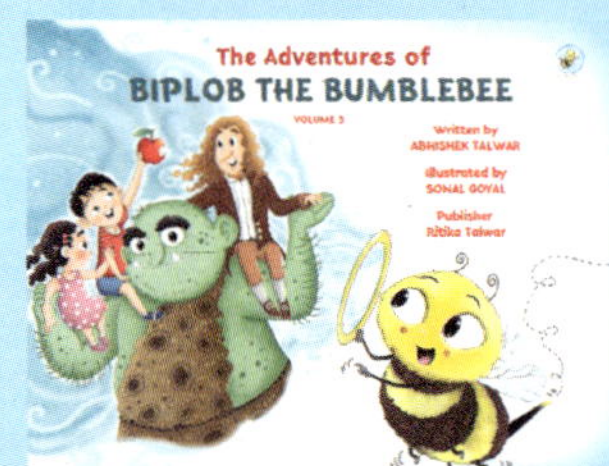
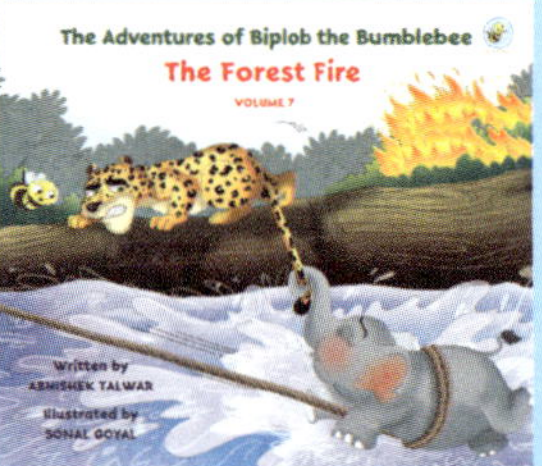
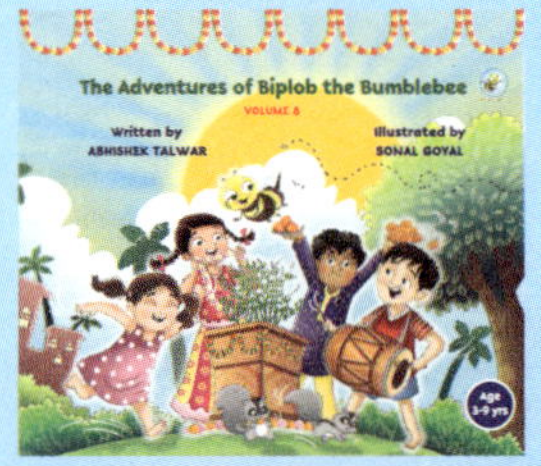

Scan to order our
other titles